Under Arctic ICE

AF621029

by

H.G. Winter

Double 9 BOOKS

Under Arctic Ice
by H.G. Winter

Copyright © 2023

All Rights reserved.

No part of this publication may be reproduced, stored in a retrieval system, or transmitted in any form or by any means, electronic, mechanical, photocopying or Otherwise, without the written permission of the publisher.
The author/editor asserts the moral right to be identified as the author/editor of this work.

ISBN: 978-93-59329-58-1

Published by

DOUBLE 9 BOOKS
2/13-B, Ansari Road
Daryaganj, New Delhi – 110002
info@double9books.com
www.double9books.com
Tel. 011-40042856

This book is under public domain

ABOUT THE AUTHOR

Hiram Gilmore "Harry" Bates III (October 9, 1900 – September 1981) was a science fiction editor and writer from the United States. The well-known science fiction film The Day the Earth Stood Still (1951) was based on his short tale "Farewell to the Master" (1940). On October 9, 1900, in Pittsburgh, Pennsylvania, Hiram Gilmore Bates III was born. In the 1920s, he began working for William Clayton as an editor of adventure pulp publications. When Clayton recommended a historical adventure magazine, Bates presented many options that he said would be easier to edit, and the result was Astounding Science Fiction. Bates, who was not a lover of science fiction, edited the journal from its start in January 1930 until its sale to Street and Smith in March 1933, when Clayton went bankrupt and the magazine was sold. During that period, he edited several other magazines for Clayton, including Strange Tales, which was meant to compete with Weird Tales. Bates thought science fiction novels were poorly written at the time: "Amazing novels! I bought a copy once. What dreadful stuff I'd discovered! Filled of trivia! Full of puerilities. Unimaginables wrote it! But now that I think about it, I wonder whether there would be a demand for a well-written magazine on Amazing subjects." Bates stated that the "science fiction of the early writers had little relation to science of the scientists."

CONTENTS

CHAPTER I
AN EMPTY ROOM 7

CHAPTER II
THE CRASH 11

CHAPTER III
THE FATE OF THE PEARY 15

CHAPTER IV
"NO CHANCE LEFT" 21

CHAPTER V
THE LAST ASSAULT 28

CHAPTER VI
IN A BISCUIT CAN 34

CHAPTER VII
THE AWAKENING 39

CHAPTER VIII
THE DUEL 43

CHAPTER I
AN EMPTY ROOM

The house where the long trail started was one of gray walls, gray rooms and gray corridors, with carpets that muffled the feet which at intervals passed along them. It was a house of silence, brooding within the high fence that shut it and the grounds from a landscape torpid under the hot sun of summer, and across which occasionally drifted the lonely, mournful whistle of a train on a nearby railroad. Inside the house there was always a hush, a heavy quiet—restful to the brain.

Ken Torrance races Poleward to the aid of the submarine *Peary*, trapped in an icy limbo of avenging sealmen.

But now a voice was raised, young, angry, impatient, in one of the gray-walled rooms.

"Yes, I rang for you. I want my bags packed. I'm leaving this minute!"

The face of the man who had entered showed surprise.

"Leaving, Mr. Torrance? Why?"

"Read this!"

As if, knowing and therefore dreading what he would see, the attendant took the newspaper held outstretched to him and followed the pointing finger to a featured column. He scanned it:

Deadline Passed for Missing Submarine

> Point Barrow, Aug. 17 (AP): Planes sent out to search for the missing polar submarine *Peary* have returned without clue to the mystery of is disappearance. The close search that has been conducted through the last two weeks, involving great risks to the pilots, has been fruitless, and authorities now hold out small hope for Captain Sallorsen, his crew and the several scientists who accompanied the daring expedition.
>
> If the *Peary*, as is generally thought, is trapped beneath the ice floes or embedded in the deep silt of the polar sea-floor, her

margin of safety has passed the deadline, it was pointed out to-day by her designers. Through special rectifiers aboard, her store of air can be kept capable of sustaining life for a theoretical period of thirty-one days. And exactly thirty-one days have now elapsed since last the *Peary's* radio was heard from a position 72° 47′ N, 162° 22′ W, some twelve hundred miles from the North Pole itself.

In official circles, hope was practically abandoned for the missing submarine, though attempts will continue to be made to locate her....

"I'm sorry, Mr. Torrance," said the attendant nervously. "This paper should—"

"Should never have reached me, eh? Through some slip of the people who censor my reading matter here, I read what I wasn't supposed to—that's what you mean?"

"It was thought better, Mr. Torrance, by the doctors, and—"

"Good God! Thought better! Through their sagacity, these doctors have probably condemned the men on this submarine to death! I haven't heard a word about the expedition; didn't even know the *Peary* was up there, much less missing!"

"Well, Mr. Torrance," the attendant stammered, more and more unsettled, "the doctors thought that—that any news about it would—well, upset you."

The young man laughed bitterly;

"Bring on my old 'trouble,' I suppose. The doctors have been considerate, but I won't concern them any more. I'm through. I'm leaving for the north—right now. There's a bare chance I might still be in time."

"I'm sorry, Mr. Torrance, but you can't."

"Can't?"

The attendant had retreated to the door. His eyes were nervous, his face pale.

"It's orders, Mr. Torrance. You've been under observation treatment, and the doctors left strict orders that you must stay."

The young man throbbed with dangerous anger. His hands clenched and unclenched. He burst out, in a last attempt at reason:

"But don't you see, I've *got* to get to the *Peary*! It's the last hope for those men! The position she was last heard from is right where I—"

"You can't leave, Mr. Torrance! I'm sorry, but I'll have to call a guard!"

For a minute their eyes held. With an effort, the young man said more calmly:

"I see. I see. I'm a prisoner. All right, leave me."

The attendant was more than willing. The young man heard the door's lock click. And then he lowered his head and pressed his hands hard into his face.

But a second later he was looking up again, at the single wide window which gave out on the lonely landscape over which sometimes came drifting the distant cry of a train's whistle.

Two months before, Kenneth Torrance had returned to the whaling submarine *Narwhal*, of which he was first torpooner, with a confused story of men who were half-seals that lived in mounds under the Arctic ice, [1] who had captured him and—he found—had also captured the second torpooner, Shanley Beddoes. In breaking free from their mound-prison, Beddoes had killed one of the sealmen and had been himself slain minutes later by a killer whale, one of the fierce scavengers of the sea which the sealmen trapped for food even as the *Narwhal* sought them for oil. Ken Torrance alone came back.

[1] See the February, 1932, issue of Astounding Stories.

Over their doubts, he had stuck to his story. Later, he had repeated it to officials of the Alaska Whaling Company, who worked the submarine and several surface ships. They in return had sent him to a private sanitarium in the State of Washington for a rest which they hoped would "iron out the kink" in his brain.

Here Ken had been for six weeks, while the exploring submarine *Peary* nosed her way northward toward the Pole. Here he had been, all unknowing, while the world hummed with reports of the *Peary's* disappearance in that far-off ever-shrouded sea of mystery.

She might, Ken knew, have struck a shaft of underwater ice, sending her to the bottom; some of her machinery might have cracked up, paralyzing her; the ice-fields under which she cruised might have shifted suddenly, crushing her ribs—of these perils the world knew as well as he. But the submarine's crew was prepared for them; the *Peary* was equipped with a circular saw for cutting up through the ice from beneath, and she carried sea-suits which would allow her men, if she were wrecked on the bottom, to leave her and get up on the ice and wait for the first searching plane.

Why, then, had not the planes which scoured the region found the survivors?

That was the mystery—but not to Ken Torrance. There was another peril, of which he alone knew. Not far from where the *Peary's* last radio report had come, a group of hollowed-out mounds lay on the sea-floor, swarming with brown-skinned, quick-swimming creatures. Sealmen, they were—men who, like the seals, had gone back to the sea. Months ago, Second Torpooner Chanley Beddoes had killed one of them. They were intelligent; they could remember; they were capable of hate and fear; they would be desirous of leveling the debt!

There, Ken felt sure, lay the reason for the *Peary's* baffling silence, for the non-appearance of her men.

There might still be time. No one of course would listen to him and believe, so he would have to go in search of the *Peary* and her crew himself.

Standing by the window, Kenneth Torrance quickly planned the several steps which would take him to the Arctic and its silent ice-coated sea.

And when, some two hours later, after a short warning rap on the door, the individual who served as Mr. Torrance's attendant entered his room, he was confronted, not by the gentleman whose dinner he carried, but by an empty room, a stripped bed, an open window, and a rope of sheets dangling from it toward the ground two stories beneath.

That was at seven o'clock in the evening.

CHAPTER II
THE CRASH

At a few minutes before eight o'clock, Air Mail Pilot Steve Chapman was enjoying a quiet cigarette while waiting for the mechanics to warm up the five hundred horses of his mail plane satisfactorily. Halfway through, he heard, from behind, a quick patter of feet, and, turning, he observed a figure clad in flannel trousers and sweater. The cigarette dropped right out of his mouth as he cried:

"Ken! Ken Torrance!"

"Thank God you're here!" said Kenneth Torrance. "I gambled on it. Steve, I've got to borrow your own personal plane."

"What?" gasped Steve Chapman. "What—what—?"

"Listen, Steve. I haven't been with the whaling company lately; been resting, down here—secluded. Didn't know that submarine, the *Peary*, was missing. I just learned. And I know damned well what's happened to it. I've got to get to it, quick is I can, and I've got to have a plane."

Steve Chapman said rather faintly:

"But—where was the *Peary* when they last heard from her?"

"Some twelve hundred miles from the Pole."

"And you want to get there in a plane? From here?"

"Must!"

"Boy, you stand about one chance in twenty!"

"Have to take it. Time's precious, Steve. I've got to stop in at the Alaska Whaling Company's outpost at Point Christensen, then right on up. I can't even begin unless I have a plane. You've got to help me on my one chance of bringing the *Peary's* men out alive! You'll probably never see the plane again, Steve, but—"

"To hell with the plane, if you come through with yourself and those men," said the pilot. "All right, kid, I don't get it all, but I'm playing with you. You're taking my own ship."

He led Ken to a hangar wherein stood a trim five-passenger amphibian; and very soon that amphibian was roaring out her deep-throated song of power on the line, itching for the air, and Steve Chapman was shouting a few last words up to the muffled figure in the enclosed control cockpit.

"Fuel'll last around forty hours," he finished. "You'll find two hundred per, easy, and twenty-five hours should take you clear to Point Christensen. I put gun and maps in the right pocket; food in that flap behind you. Go to it, Ken!"

Ken Torrance gripped the hand outstretched to his and held it tight. He could say nothing, could only nod—this was a real friend. He gave the ship the gun.

Her mighty Diesel bellowed, lashed the air down and under; the amphibian spun her retractable wheels over the straight hard ground until they lifted lightly and tilted upward in a slow climb for altitude. With fiery streams from the exhaust lashing her flanks, she faded into the darkness to the north.

"Well," murmured Steve Chapman, "I've got her instalments left, anyway!" And he grinned and turned to the mail.

That night passed slowly by; and the next day; and all through night and day the steady roar of beating cylinders hung in Kenneth Torrance's ears. At last came Point Christensen and a descent; sleep and then quick, decisive action; and again the amphibian rose, heavily loaded now, and droned on toward the ice and the cold bleak skies of the far north. On, ever on, until Point Barrow, Alaska's northernmost spur, was left behind to the east, and the world was one of drifting ice on gray water. Muscles cramped, mind dulled by the everlasting roar, head aching and weary, Ken held the amphibian to her steady course, until a sudden wind shook her momentarily from it.

A rising wind. The skies were ugly. And then he remembered that the men at Point Christensen had warned him of a storm that was brewing. They'd told him that he was heading into disaster; and their surprised, rather fearful faces appeared before him again, as he had seen them just before taking off, after he had told them where he was going.

Of course they'd thought him crazy. He had brought the amphibian down in the little harbor off the whaling company's base, gone ashore and greeted his old friends. There was only a handful of men stationed there; the *Narwhal* was being overhauled in a shipyard at San Francisco, and it wasn't the season for surface whalers. They knew that he, Ken, had been put in a sanitarium; all of them had heard his wild story about sealmen. But he concocted a plausible yarn to account for his arrival, and they had fed him and given him a berth in the bunkhouse for the night.

For the night! Ken Torrance grinned as he recalled the scene. In the middle of the night he had risen, quickly awakened four of the sleeping men, and with his gun forced them to take a torpoon from the outpost's storehouse and put it inside the amphibian's passenger compartment.

It was robbery, and of course they'd thought him insane, but they didn't dare cross him. He had told them cheerfully he was going after the *Peary*, and that if they wanted the torpoon back they were to direct the searching planes to keep their eyes on the place where the submarine was last heard from....

Ken came back to the present abruptly as the plane lurched. The wind was getting nasty. At least he did not have much farther to go; an hour's flying time would take him to his goal, where he must descend into the water to continue his search. His search! Had it been, he wondered, a useless one from the start? Had the submarine's crew been killed before he'd even read of her disappearance? If the sealmen got them, would they destroy them immediately?

"I doubt it," Ken muttered to himself. "They'd be kept prisoners in one of those mounds, like I was. That is, if they haven't killed any of the creatures. It hangs on that!"

An hour's time, he had reckoned; but it was more than an hour. For soon the world was blotted out by a howling dervish of wind and driven snow that time and time again snatched the amphibian from Ken's control and hurled it high, or threw it down like a toy toward the inferno of sea and ice he knew lay beneath. He fought for altitude, for direction, pitched from side to side, tumbled forward and back, gaining a few hundred feet only to feel them plucked breathtakingly out from under him as the screaming wind played with him.

Now and again he snatched a glance at the torpoon behind. The gleaming, twelve-foot, cigar-shaped craft, with its directional rudders, propeller, vision-plate and nitro-shell gun lay safely secured in the passenger compartment, a familiar and reassuring sight to Ken, who, as first torpooner of the *Narwhal*, had worked one for years in the chase for killer whales. Soon, it seemed, he would have to depend on it for his life.

For all the Diesel's power, it was not enough to cope with the dead weight of ice which was forming over the plane's wings and fuselage. He could not keep the altimeter up. However he fought, Ken saw that finger drop down, down—up a trifle, quivering as the racked plane quivered—and then down and down some more.

He saw that the plane was doomed. He would have to abandon it—in the torpoon—if he could.

He was some thirty miles from his objective. The sea beneath would be half hidden under ragged, drifting floes. In fair weather he could have chosen a landing space of clear water, but now he could not choose. The altitude dial said that the water was three hundred feet beneath, and rapidly rising nearer.

A margin of seconds in which to prepare! Ken locked the controls and scrambled back into the passenger compartment. Steadying himself on the bucking floor, he opened the torpoon's entrance port and slid in; quickly he locked the port and strapped the inner body harness around him; and then he waited.

Now it was all chance. If the plane crashed into clear water, he was safe; but if she hit ice.... He put that thought from him.

The locked controls held the amphibian for perhaps thirty seconds. Then with a scream the storm-giant took her. A mad up-current of wind hurled her high, whirled her dizzily, toyed with her—and then she spun and dove. Down, down, down; down with a speed so wild Ken grew faint; down through the core of a maelstrom of snow till she crashed.

Kenneth Torrance knew a sudden shaking impact; for an instant there was uncertainty; and then came all-pervading quiet....

CHAPTER III
THE FATE OF THE PEARY

Quiet, and utter, liquid darkness.

Liquid! Around him, Ken heard a gurgling, at first loud and close, then subsiding to a low whispering of currents. The amphibian had hit water.

Gone in an instant was the shriek and fury of the storm and in its place the calm, slow-heaving silence of underwater. The plane was shattered in a dozen places, but the torpoon had easily stood it.

Ken turned to action. He switched on the torpoon's dashboard lights and twin bow-beams, and saw that the shell was wedged in the fuselage. The plane was apparently entirely under the surface, and her interior filled with water.

Holding the propeller in neutral, he revved up the powerful electric motor. Then he bit the propeller in, slowly. The torpoon nudged back for inches. Then, throwing the gear into forward, Ken gave her full speed. The torpoon leaped ahead, crunched through the weakened corner ahead and was free.

It was a world of drab tones that she came into. Down below was impenetrable blackness, shading softly overhead into blue-gray which was mottled by lighter areas from breaks in the floes above. All was calm. There was no sign of life save for an occasional vague shadow that, melting swiftly away, might have been a fish or seaweed. Placid always, would be this shrouded sea of mystery, no matter what furious tempest raged above over the flat leagues of ice and water.

But the seeming peacefulness was but a mask for danger. Kenneth Torrance's face was set in sober lines as he sped the slim torpoon northward, her bow lights shafting long white fingers before her. For now there was only one path—and that lay ahead. He could not turn back. Storm and water had destroyed the plane that could take him back to land. He could not possibly reach any outpost of civilization in the torpoon, for her cruising radius was only twenty hours. He had planned to land the amphibian on the ice above the spot where the *Peary* had disappeared, then find a break in the ice and slide down below in the torpoon on his quest—to return to the plane if it proved fruitless. But now there was no retreat. It was succeed, or die.

And with that realization a more dreadful thought flashed into his mind. All those men, of the whaling company and the sanitarium, thought him a little crazy. And, since lunatics are always convinced of the reality of their visions, what if the sealmen—his adventure amidst them—had been but a dream, a nightmare, an hallucination? What if he were in truth crazy? The fear grew rapidly. What if he were? God! He, hunting for the *Peary*, when all those planes and men had failed! He, expecting to achieve what those searchers, with far greater resources, had not been able to! Did not that give evidence that his mind was twisted? Creatures, half-seal, half-men, living under the ice—it certainly seemed a lunatic's obsession.

Then something within him rose and fought back.

"No!" he cried aloud. "I'll go bugs if I think like that! Those sealmen were real—and I know where they are. I'm going on!"

And, an hour later, the dashboard's shaded dials told him he was on the exact spot where the *Peary* had last reported....

Here was the real Arctic, the real polar sea. No sun, no breath of the world above could reach it through its eternal mask of solid ice. As one of the few unfamiliar aspects of the earth, it was as far removed from the imagination of man as if it were part of a far planet hung spinning millions of miles out in space. Men could reach it in shells of metal, but it was not meant for him, and was always hostile. A dozen times a daring one could cross safely its cold lonely reaches, but the thirteenth time it would snare and destroy him for the unwanted trespasser he was.

It was here that the *Peary* had stepped off into mystery. At this point her hull had throbbed with air, movement, life; at this point all had been well. And then, minutes or hours later, close to here, the sea devil had sprung.

What had happened? What had trapped her? What, even more baffling, had kept her men with their manifold safety devices from even reaching and climbing up on the ice above to signal the searching planes?

Ken Torrance, oppressively alone in the hovering torpoon, gazed through its vision-plate of fused quartz around him. Gray sea, filtering to black beneath; distant eerie shadows, probably meaning nothing, but possibly all important; ceiling of thick ice above, rough and in places broken by a sharp down-thrusting spur—these were his surroundings. These were what he must hunt through, until he came upon the crumpled remnant of a submarine, or the murky, rounded hillocks which gave habitation to the creatures he suspected of capturing that submarine's crew.

He began the search systematically. He angled the torpoon down to a position halfway between sea-floor and ice-ceiling, then swung her in an ever-widening circle. Soon his orbit had a diameter of a half-mile; then a mile; then two.

The torpoon slipped through the water at full speed, her light-beams like restless antennae, now stabbing to the right to dissolve a formless shadow, now to the left to throw into blinding white relief a school of half-transparent fish which scurried with frantic wrigglings of tails from the glare, now slanting up to bathe the cold glassy face of an inverted ice-hill, now down to dig two white holes in the deeper gloom.

Ken continued this routine for hours. Steadily and low the electric motor droned in the ears of the watchful pilot, and the stubby propeller's blades flashed round in a blur of speed between the slightly slanted rudders. Somewhere, miles away, a splintered amphibian plane was slipping down to her last landing, and above, perhaps, the white hell of storm which had brought her low still bowled over the trackless wastes; but here were only shadows and shifting gloom, straining the alert eyes to soreness and tensing the watcher's brain with alarms that, one after another, were only false.

Until at last he found her.

Immediately he shut off all his lights. He no longer needed them. Far in the distance, and below, wavered a faint yellow glow. It was no fish; it could mean only one thing—the lights of a submarine.

And lights meant life! There would be none burning in a deserted submarine. His heart beat fast and his tight, sober lips widened in a quick grin. He had found the *Peary*! And found her with some life still aboard her! He was in time!

So Ken rejoiced while he slid the torpoon down to a level just a few feet above the silty sea bottom, reducing her to quarter-speed. There was an urge inside him to switch on his bow-beams, reach them out toward the submarine's hull to tell all within that help was at last at hand; he wanted to send the torpoon ahead at full speed. But caution restrained him to a more deliberate course. He was in the realm of the sealmen, and he did not wish to attract the attention of any. So he advanced like a furtive shadow slinking along the dark sea-bottom, deep in the covering gloom.

Nearer and nearer, while the distant blur of yellow light grew. Nearer and nearer to the long-trapped men, while the consciousness that he had succeeded intoxicated him. He alone had found them! Sealmen or no sealmen, he had found the *Peary*! And found her with lights lit and life inside! Nearer and nearer....

And then suddenly Ken halted the torpoon and stared with wide, alarmed eyes. For the submarine was now plainly visible in detail—and he saw her real plight and with it knew the answer to the mystery of her long silence and the non-appearance of her men on the ice field above.

The *Peary* was a spectacle of fantastic beauty. It was as if a huge, rounded piece of amber, mellow, golden, lay in the murk of the sea-floor. Not steel, hard and grim, but of transparent, shimmering stuff she was built, all coated a soft yellow by her lights, clearly visible inside. Ken had known something of her radical construction; knew that a substance called quarsteel, similar to glass and yet fully as tough as steel, had been used for her hull, making her a perfect vehicle for undersea exploration. Her bow was capped with steel, and her stern, propellers, diving rudders; her port-locks, for the releasing of torpoons, were also of steel, as were the struts that braced her throughout—but the rest was quarsteel, glowing and golden as the heart of amber.

Beautiful with a wild yet scientific beauty was the *Peary*, but she was not free. She was trapped. She was fastened to the mud of the gloomy sea-floor.

Ropes held her down; and Ken Torrance knew those ropes of old. They were tough and strong, woven of many strands of seaweed, and twenty or thirty of them striped the *Peary's* two hundred feet of hull. Unevenly spaced, stretched clear over the ship from one side to the other, they were caught around her up-jutting conning tower, fastened through her rudders, and holding tight in a score of places. They held the submarine down despite all the buoyancy of her emptied tanks and the power of her twin propellers.

And the sealmen swam around her.

Restless dark shadows against the golden hull, they wavered and darted and poised, totally unafraid. Another in Kenneth Torrance's place would have put them down as some strange school of large seals, inordinately curious but nothing more; but the torpooner knew them as men—men remodeled into the shape of seals; men who, ages ago, had forsaken the land for the old home of all life, the sea; who, through the years, had gradually changed in appearance as their flesh had become coated with layers of cold-resisting blubber; whose movements had become adapted to the water; whose legs and arms had evolved into flippers; but whose heads still harbored the now faint spark of intelligence that marked them definitely as men.

Emotions similar to man's they had, though dulled; friendliness, curiosity, anger, hate, and—Ken knew and feared—even a capacity for vengeance. Vengeance! An eye for an eye, tooth for a tooth—the old law peculiar to man! Chanley Beddoes had slain one of them; if only the *Peary's* crew had not killed more! If only that, there might be hope!

First he must get inside the submarine. Warily, like a stalking cat, Ken Torrance inched the torpoon toward the great shining ship. At least he was in time. Within her he could see figures, most of them stretched out on the decks of her different compartments, but one of whom occasionally moved—slowly. He understood that. For weeks now the *Peary* had lain captive, and her air had passed beyond the aid of rectifiers. Tortured, those survivors inside were, constantly struggling for life, with vitality ever sinking lower. Some might already be dead. But at least he could try to save the rest.

He approached her from one side of the rear, for in the rear compartment were her two torpoon port-locks. The one on his side was empty, its outer door open. The torpoon it had held had been sent out, probably for help, and had not returned. It provided a means of entrance for him.

At perhaps a hundred feet from the port-lock, Ken halted again. His slim craft was almost indistinguishable in the murk: he felt reasonably safe from discovery. For minutes he watched the swimming sealmen, waiting for the best chance to dart in.

It was then, while studying the full length of the submarine more closely, that he saw that one compartment of her four was filled with water. Her steel-caped bow had been stove in. That, he conjectured, had been the original accident which had brought her down. It was not a fatal accident in itself, for there were three other compartments, all separated by watertight bulkheads, and the flooded one could be repaired by men in sea-suits—but then the sealmen had come and roped her down where she lay. Some of the creatures, he saw, were actually at that time inside the bow compartment, swimming around curiously amidst the clustered pipes, wheels and levers. It was a weird sight, and one that held his eyes fascinated.

But suddenly, through his absorption, danger prickled the short hairs of his neck. A lithe, sinuous shadow close ahead was wavering, and large, placid brown eyes were staring at him. A sealman! He was discovered! And instinctively, immediately, Ken Torrence brought the torpoon's accelerator down flat.

The shell jumped ahead with whirling propeller. The creature that had seen him doubled around and sped in retreat. In brief snatches, as the torpoon streaked across the hundred-foot gap to the empty port-lock, Ken glimpsed his discoverer gathering a group of its fellows, and saw brown-skinned bodies swarm after him with nooses of seaweed-rope—and then the great transparent side wall of the *Peary* was before him, and the port-locks dark opening. Ken threw his motor into reverse, slid the torpoon slightly to one side, and there was a jerk, a jar, and a sensation of something moving behind.

He turned to see the port-lock's outer door closing, activated by controls inside the submarine—and just in time to shut out the first of his pursuers. Then the port-lock's pumps were draining the water from the chamber, and the inner door clicked and opened.

Kenneth Torrance climbed stiffly from the torpoon to enter the interior of the long-lost and besieged exploring submarine *Peary.*

CHAPTER IV
"NO CHANCE LEFT"

His entrance was an unpleasant experience. He had forgotten the condition of the air inside the submarine, and what its effect on him, coming straight from comparatively good and fresh air, would be, until he was seized by a sudden choking grip around his throat. He reeled and gasped, and was for a minute nauseated. Lights flashed around him, and teetering backward he leaned weakly, against some metal object until gradually his head cleared; but his lungs remained tortured, and his breathing a thing of quick, agonised gulps.

Then came sounds. Figures appeared before him.

"From where—" "Who are you?"

"What—what—what—" "How did you?"

The half-coherent questions were couched in whispers. The men around him were blear-eyed and haggard-faced, their skins dry and bluish, and not a one was clad in more than undershirt and trousers. Alive and breathing, they were—but breathing grotesquely, horribly. They made awful noises at it; they panted, in quick, shallow sucks. Some lay on the deck at his feet, outstretched without energy enough to attempt to rise.

Beautiful and slumber-like the submarine had appeared from outside, but inside that effect was lost. There were the usual appurtenances: a maze of pipes, wheels, machinery, all silent now, and cold; here were the two port-locks for torpoons; the emergency steering controls; the small staterooms of the *Peary's* officers. Looking forward, still striving for complete clear-headedness and normality, Ken could see the two intact forward compartments, silent and apparently lifeless, with dim lamps burning. They ended with the watertight bulkhead which stood between them and the flooded bow compartment.

Ken at last found words, but even his short query cost a sickening effort.

"Where's—the commander?" he asked.

A man turned from where he had been leaning against a nearby wheel control. He was stripped to the waist. His tall body was stooped, and the skin of his ruggedly cut face drawn and parchment-like. His face had once been dignified and authoritative, but now it was that of a man who nears death after a long, bitter fight for life. The smile which he gave to Ken was painful—a mockery.

"I am," he said faintly. "Sallorsen. Just wait, please. A minute. I worked port-lock. Breath's gone...."

He sucked shallowly for air and let his smile go. And standing there, beside him, gazing at the worn frame, Ken felt strength come back. He had just entered; this man and the others had been here for weeks!

"I'm Sallorsen," the captain went on at last. All his words were clipped off, to cost minimum effort. "Glad you got through. Afraid you're come to prison, though."

"No!" Ken said emphatically. He spoke to the captain, but what he said was also for all the others grouped around him. "No, Captain! I'm Kenneth Torrance. Once torpooner with Alaska Whaling Company. They thought me crazy—crazy—'cause I told about sealmen. Put me in sanitarium. I knew they had you—when—heard you were missing." He pointed at the brown-skinned creatures that clustered close around the submarine outside her transparent walls. "I got free and came. Just in time."

"In time? For what?"

Another voice gasped out the question. Ken turned to a broad-shouldered man with a ragged growth of beard that had been a trim Van Dyke; and before the torpooner could answer, Sallorsen said:

"Dr. Lawson. One of our scientists. In time for what?"

"To get you and the submarine free," said Ken.

"How?"

Ken paused before replying. He gazed around—out the side walls of glistening quarsteel into the sea gloom, into the thick of the smooth, lithe, brown-skinned shapes that now and again poised pressing against the submarine, peering in with their liquid seal's eyes. Dimly he could see the taut seaweed ropes stretching down from the top of the *Peary* to the sea-bottom. It looked hopeless, and to these men inside it was hopeless. He knew he must speak in confident, assured tones to drive away the uncaring lethargy holding them all, and he framed definite, concise words with which to do it.

"These creatures have caught you," he began, "and you think they want to kill you. But look at them. They seem to be seals. They're not. They're men! Not men like us—half-men—sealmen, rather—changed into present form by ages of living in the water. I know. I was captured by them once. They're not senseless brutes; they have a streak of man's intelligence. We must communicate with that intelligence. Must reason with them. I did once. I can do it again.

"They're not really hostile. They're naturally peaceful; friendly. But my friend—dead now—killed one of them. Naturally they now think all creatures like us enemies. That's why they trapped your sub.

"They think you're enemies; think you want to kill them. But I'll tell them—through pictures, as I did once before—that you mean them no harm. I'll tell them you're dying and must have air—just as they must. I'll tell them to release submarine and we'll go away and not disturb them again. Above all I must get across that you wish them no harm. They'll listen to what my pictures will say—and let us go—'cause at heart they're friendly!"

He paused—and with a ghastly, twisted smile, Captain Sallorsen whispered:

"The hell you say!"

His sardonic comment brought a sudden chill to Kenneth Torrance. He feared one thing that would render his whole value useless. He asked quickly:

"What have you done?"

"Those seals," Sallorsen's labored voice continued "—they've killed eight of us. Now they're killing all."

"But have you killed any of them?" Breathless, Ken waited for the answer be feared.

"Yes. Two."

The men were all staring at Ken, so he had to hide the awful dejection which clamped his heart. He only said:

"That's what I feared. It changes everything. No use trying to reason with them now." He fell silent. "Well," he said at last, trying to appear more cheerful, "tell me what happened. Maybe there's something you've overlooked."

"Yes," Sallorsen whispered. He started to come forward to the torpooner, but stumbled and would have fallen had not Ken caught him in time. He put one of the captain's arms around his shoulder, and one of his own around the man's waist.

"Thanks," Sallorsen said wryly. "Walk forward. Show you what happened."

There were men in the second compartment, and they still fought to live. From the narrow seamen's berths that lined the walls came the sound of breathing even more torturous than that of the men in the rear. In the single bulb's dim light Ken could see their shapes stretched motionlessly out, panting and panting. Occasionally hands reached up to claw at straining necks, as if to try and rid throats of strangling grasps. Two figures had won free from the long struggle. They lay silent and still, the outline of their dead bodies showing through the sheets pulled over them.

Slowly Sallorsen led Ken through this compartment and into the next, which was bare of men. Here were the ship's main controls—her helm, her central multitude of dials, levers and wheels, her televisiscreen and old-fashioned emergency periscope. A metal labyrinth it was, all long silent and inactive. Again the weird contrast struck Ken, for outside he could still see the scene of vigorous, curious life that the sealmen constituted. Close they came to the submarine's sheer walls of quarsteel, peering in stolidly, then flashing away with an effortless thrust of flippers, sometimes for air from some break in the surface ice.

Like men, the sealmen needed air to live, and got it fresh and clean from the world above. Inside, real men were gasping, fighting, hopelessly, yielding slowly to the invisible death that lay in the poisonous stuff they had to breathe....

Ken felt Sallorsen nudge him. They had come to the forward end of the control compartment, and could go no farther. Before them was the watertight door, in which was set a large pane of quarsteel. The captain wanted him to look through.

Ken did so, knowing what to expect; but even so he was surprised by the strangeness of the scene. In among the manifold devices of the front compartment, its wheels and pipes and levers, glided slowly the sleek, blubbery shapes of half a dozen sealmen. Back and forth they swam, inspecting everything curiously, unhurried and unafraid; and as Ken stared one of them came right up to the other side of the closed watertight door, pressed close to the pane and regarded him with large placid eyes.

Other sealmen entered through a jagged rip in the plates on the starboard side of the bow. At this Sallorsen began to speak again in the short, clipped sentences, punctuated by quick gasps for air.

"Crashed, bow-on," he said. "Underwater ice. Outer and inner plates crumpled like paper. Lost trim and hit bottom. Got this door closed, but lost four men in bow compartment. Drowned. No chance. Sparks among 'em, at his radio. That's why we couldn't radio for help." He paused, gasping shallowly.

"Could've got away if we'd left immediately. One flooded compartment not enough to hold this ship down. But I didn't know. I sent two men out in sea-suits—inspect damage. Those devils got them.

"The seal-things came in a swarm. God! Fast! We didn't realize. They had ropes, and in seconds they'd lashed us down to the sea-floor. Lashed us fast!" Again he paused and sucked for the poisoned air, and Ken Torrance did not try to hurry him, but stood silent, looking forward to the squashed bow, and out the sides to where he could see the taut black lines of the seaweed-ropes.

"The two men put up fight. Had crowbars. Useless—but they killed one of the devils. That did it. They were torn apart in front of us. Ripped. Mangled. By spears the things carry. Dead like that."

"Yes," murmured Ken, "that would do it...."

"I quick tried to get away," gasped Sallorsen. "Full-speed—back and forth. No good. Ropes held. Couldn't break. All our power couldn't! So then—then I acted foolishly. Damn foolish. But we were all a little crazy. A nightmare, you know. Couldn't believe our eyes—those seals outside, mocking us. So I called for volunteers. Four men. Put 'em in sea-suits, gave 'em shears and grappling prongs. They went out.

"They went out laughing—saying they'd soon have us free! Oh, God!" It seemed he could not go on, but he forced the words out deliberately. "Killed without a chance! Ripped apart like the others! No chance! Suicide!"

Ken felt the agony in the man, and was silent for a while before quietly asking:

"Did they kill any more of the sealmen?"

"One. Just one. That made two of them—six of us. What the hell are the rest of them waiting for?" Sallorsen cried. "They killed eight in all! To our two! That's enough for them, isn't it?"

"I'm afraid not," said Ken Torrance. "Well, what then?"

"Sat down and thought. Carefully. Hit on a plan. Took one of our two torpoons. Lashed on it steel plates, ground to sharp cutting edges. Spent days at it. Thought torpoon could go out and cut the ropes. Haines volunteered and we shot him and torpoon out."

"They got the torpoon?" Ken asked.

Sallorsen's arm raised in a pointing gesture. "Look."

Some fifty feet away from the *Peary*, on the side opposite to the one Ken Torrance had approached, a dimly discernible object lay in the mud.

In miniature, it resembled the submarine: a cigar-shaped steel shell, held down to the sea-bottom by ropes bound over it. Cutting edges of steel had been fastened along its length.

"I see," said Ken slowly. "And its pilot?"

"Stayed in the torpoon thirty-six hours. Then went crazy. Put on sea-suit and tried to get back here. Whisk—they got him. Killed and mangled while we watched!"

"But didn't his torpoon have a nitro-shell gun? Couldn't he have fought them off for a time?"

"Exploring submarine, this! No guns in torpoons like whalers. Gun wouldn't help, anyway. These devils too fast. No use. No hope anywhere...." Sallorsen sank back against the bulkhead, his lips moving but no sound coming forth. Dully he stared ahead, through the submarine, for a moment before uttering a cackling mockery of a laugh and going on.

"Even after that, still hoped! Blew every tank on ship; blew out most of her oil. Threw out everything not vital. Lightened her as much as could. Machinery—detachable metal—fixtures—baggage—instruments—knives, plates, cups—everything! She rose a couple of feet—no more! Put motors at full speed—back and forth—again, again, again. Buoyancy—power—no good. No damn good!

"And then we tried the last chance. Explosives. Had quite a store, Nitromite, packed in cases; time-fuses to set it off. Had it for blasting ice. I sent up a charge and blew hole in the ice overhead, for our other torpoon.

"Nothing else left. Knew planes must be nearby, searching. Last torpoon was to shoot up to the hole—pilot to climb on ice and stay there to signal a plane."

"Did he get there?"

"Hell no!" Sallorsen cackled again. "It was roped like the other. Pilot tried to get back, but they got him like first. There's the torpoon—out ahead."

Ken could just make it out. It lay ahead, slightly to port, lashed down like its fellow by seaweed-ropes. His eyes were held by it, even when Sallorsen continued, in an almost hysterical voice:

"Since then—since then—you know. Week after week. Air getting worse. Rectifiers running down. No night, no day. Just the lights, and those damned devils outside. Wore sea-suits for a while; used twenty-nine of their thirty hours air-units. Old Professor Halloway died, and another man. Couldn't do anything for 'em. Just sit and watch. Head aching, throat choking—God!...

"Some of the men went mad. Tried to break out. Had to show gun. Quick death outside. Here, slow death, but always the chance that—Chance, hell! There's no chance left! Just this poison that used to be air, and those things outside, watching, watching, waiting—waiting for us to leave—waiting to get us all! Waiting...."

"Something's up!" said Ken Torrance suddenly. "They've got tired of waiting!"

CHAPTER V
THE LAST ASSAULT

Sallorsen turned his head and followed the torpooner's intent, amazed gaze.

Ken said:

"There's proof of their intelligence! I've been watching—didn't realize at first. Look, here it comes!"

Several sealmen, while Sallorsen had been talking, had come dropping down from the main mass of the horde, and had grouped around the abandoned torpoon which lay some feet ahead of the submarine's bow. Expertly they had loosened the seaweed-ropes which bound it to the sea-floor, then slid back, watching alertly, as if expecting the torpoon to speed away of its own accord. Its batteries, of course, had worn out weeks before, so the steel shell did net budge. The sealmen came down close to it again, and lifted it.

They lifted it easily with their prehensile flipper-arms, and with maneuvering of delicate sureness guided it through the gash in the *Peary's* bow. Inside, they hesitated with it, midway between deck and ceiling of the flooded compartment. They poised for perhaps a full minute, judging the distance, while the two men stared; and then quickly their powerful tail flippers lashed out and the torpoon jumped ahead. It sped straight through the water, to crash its tough nose of steel squarely into the quarsteel pane of the watertight door, then rebounded, and fell to the deck.

"My God!" gasped Sallorsen. But Ken wasted no words then. He pressed closer to the quarsteel and examined it minutely. The substance showed no visible effect, but the action of the sealmen destroyed whatever hope he had felt.

The sealmen had swerved aside at the last minute; and now, picking up the torpoon again and guiding it back to the other end of the compartment, they hurled it once more with a resounding crash into the quarsteel pane.

"How long will it last under that?" Ken asked tersely.

Obviously, Sallorsen's wits were muddled at this turn. He remained gaping at the creatures and at the torpoon, now turned against its mother submarine. Ken repeated the question.

"How long? Who knows? It's as strong as steel, but—there's the pressure—and those blows hit one spot. Not—long."

Capping his words, there re-echoed again the loud crash of the torpoon's on the quarsteel. The sealmen were working in quick routine now; back and quickly forward, and then the crash and the reverberation; and again and again....

The ominous crash and ringing echoes regularly repeated, seemed to disorganise Ken's mind as he looked vainly for something with which to brace the door. Nothing unattached was left—nothing! He ran and examined the quarsteel pane again, and this time his brain heated in alarm. A thin line had shot through the quarsteel—the beginning of a crack.

"Back!" Ken shouted to the still staring Sallorsen. "Back to the third compartment. This door's going!"

"Yes," Sallorsen mumbled. "It'll go. So will the others. They'll smash them all. And when this is flooded—no hope of running the submarine again. Controls in here."

"That's too damned bad!" Ken said roughly. "Are there any sea-suits, food, supplies in here?"

"Only food. In those lockers."

"I'll take it. Get into that third compartment—hear me?" ordered Kenneth Torrance. "And have its door ready to close!"

He shoved Sallorsen away, opened the indicated lockers and piled his arms with the tins revealed. He had time for no more than one load. He jumped back into the third compartment of the *Peary* just as a splintering crash sounded from behind. The door between was swung closed and locked just as the one being battered crashed inward.

Turning, Ken saw that the torpoon had cracked through the weakened quarsteel and tumbled in a mad cascade of water to the deck of the abandoned second compartment. In dread silence, he, with Sallorsen and those of the men who had strength and curiosity enough to come forward, watched the compartment rapidly fill—watched until they saw the water pressed high against the door. And then horror swept over Ken Torrance.

Water! There was a trickle of water down the quarsteel he was leaning against! A fault along the hinge of the door—either its construction, or because it had not been closed properly.

Ken pointed it out to the captain.

"Look!" he said. "A leak already—just from the pressure! This door won't last more than a couple of minutes when they start on it—"

Sallorsen stared stupidly. As for the rest; Ken might not have spoken. They were as if in a trance, watching dumbly, with lungs automatically gasping for air.

One of the seal-creatures eeled through the shattered quarsteel of the first door and swam slowly around the newly flooded compartment. At once it was joined by five other lithe, sleek shapes which, with placid, liquid eyes, inspected the compartment minutely. They came in a group right up to the next door that barred their way and, with no visible emotion, stared through the quarsteel pane at the humans who stared at them. And then they gracefully turned and slid to the battered torpoon.

"Back!" Ken shouted, "You men!" He shook them, shoved them roughly back toward the fourth, and last, compartment. Weakly, like automatons they shuffled into it. The torpooner said bruskly to Sallorsen:

"Carry those tins of food back. Hurry! Is there anything stored in here we'll need? Sallorsen! Captain! Is there anything—"

The captain looked at him dully; then, understanding, a cackle came from his throat. "Don't need anything. This is the end. Last compartment. Finish!"

"Snap out of it!" Ken cried. "Come on, Sallorsen—there's a chance yet. Is there anything we'll need in here?"

"Sea-suits—in those lockers."

Ken Torrance swung around and rapidly opened the lockers. Pulling out the bulky suits, he cried:

"You carry that food back. Then come and help me."

But of the corner of his eye, as he worked, he could see the ominous preparations beyond in the flooded compartment—the sealmen raising the torpoon, guiding it back to the far end; leveling it out. Ken was sure the door could not stand more than two or three blows at the most. Two or three minutes, that meant—but all the sea-suits had to go back into the fourth compartment!

He was in torment as he worked. For him, the conditions were just as bad as for the men who had lived below in the submarine for a month; the poisonous, foul air racked him just as much; what breath he got he fought for just as painfully. But in his body was a greater store of strength, and fresher muscles; and he taxed his body to its very limit.

Panting, his head seeming on the point of splitting, Ken Torrance stumbled through into the last compartment laden with a pile of sea-suits. He dropped them clattering in a pile around his feet and forced himself back again. Another trip; and another....

It would never have been done had not Sallorsen and Lawson, the scientist, come to his aid. The help they offered was meager, and slow, but it sufficed. Laden for the fifth time, Ken heard what he had been anticipating for every second of the all too short, agonizing minutes: a sharp, grinding crack, and the following reverberation. He snatched a glance around to see the torpoon falling to the deck of the second compartment—the sealmen lifting it swiftly again—and a thin but definite sliver in the quarsteel of the door.

But the last suit was gotten into the fourth compartment, and the connecting door closed and carefully locked and bolted. The removal of the suits, had been achieved—but what now?

Panting, completely exhausted, Ken forced his brain to the question. From every side he attacked the problem, but nowhere could he find the loophole he sought. Everything, it seemed, had been tried, and had failed, during the *Peary's* long captivity. There was nothing left. True, he had his torpoon, and its nitro-shell gun with a clip of nineteen shells; but what use were shells? Even if each one accounted for one of the sealmen, there would still remain a swarm.

And the sea-suits. He had struggled for them and had saved them, but what use could he put them to? Go out leading a desperate final sally for the hole in the ice above? Death in minutes!

No hope. Nothing. Not even a fighting chance. These seal-creatures, strange seed of the Arctic ice, had trapped the *Peary* all too well. On the roll of mysteriously missing ships would her name go down; and he, Ken Torrance, would be considered a lunatic who had sought suicide, and found it....

Of the twenty-one survivors of the *Peary's* officers and crew, only a dozen had the will to watch the inexorable advance of the sealmen. The rest lay in various attitudes on the deck of the rear compartment, showing no sign of life save torturous, shallow pantings for air and, occasionally, spasmodic clutchings at their throats and chests, as they tried to fight off the deadly, invisible foe that was slowly strangling them.

Ken Torrance, Sallorsen, the scientist, Lawson, and a few others were pressed together at the last watertight door, peering through the quarsteel at the sea-creatures' systematic assault on the door leading into the third compartment. A straight, hard smash at it; another final splintering smash—and again the torpoon pushed through in the van of a cascade of icy, greenish water, which quickly claimed the control compartment for the attackers behind. The creatures were growing bolder. More and more of them had entered the submarine, and soon each open compartment was filled from deck to ceiling with the slowly turning, graceful brown bodies, inspecting minutely the countless wheels and levers and gauges, and inspecting also, in turns, the pale, worn faces that stared with dull eyes at them through the sole remaining door.

There was no further retreat, now. Behind was only water and the swarm that passed to and fro through it. Water and sealmen—ahead, above, to the sides, behind—everywhere. Cooped in their transparent cell, the crew of the submarine *Peary* waited the end.

Once more, as well as he could with his throbbing head and heavy, choking body, Kenneth Torrance tracked over the old road that had brought him nowhere, but was the only road open. Carefully he took stock of everything he had that he might possibly fight with.

There were sea-suits for the men, and in each suit an hour's supply of artificial but invigorating air. Two port-locks, one on each side of the stern compartment. A torpoon, with a gun and nineteen shells. Nothing else? There seemed to be, in his mind, a vague memory of something else ... something that might possibly be of use ... something.... But he could not remember. Again and again the agony of slow strangulation he was going through drove everything but the consciousness of pain from his shirking mind. But there was something else—and perhaps it was the key. Perhaps if he could only remember it—whatever it was—whether a tangible thing or merely a passing idea of hours ago—the way out would be suddenly revealed.

But he could not remember. He had the sea-suits, the port-locks and the torpoon: what possible pattern could he weave them into to bring deliverance?

No, there was nothing. Not even a girder that could be unfastened in time to brace the last door. No way of prolonging this last stand!

Beside Ken, the strained, panting voice of Lawson whispered:

"Getting ready. Over soon now. All over."

All save five of the sealmen had left the third compartment, to join the swarm constantly swimming around and over the submarine outside. The five remaining were the crew for the battering ram. With measured and deliberate movements they ranged their lithe bodies beside the torpoon, lifted it and bore it smoothly back to the far end of the compartment. There they poised for a minute, while from the men watching sounded a pathetic sigh of anticipation.

As one, the five seal-creatures lunged forward with their burden.

Crash! And the following dull reverberation.

The last assault had begun.

CHAPTER VI
IN A BISCUIT CAN

Ken Torrance glanced with dull, hopeless eyes over the compartment he stood in. Figures stretched out all over the deck, gasping, panting, strangling—men waiting in agony for death. His head sank down, and he wiped wet hands across his aching forehead. Nothing to do but wait—wait for the end—wait as the patient horde outside had been waiting in the sea-gloom for their moment of triumph, when the soft bodies inside the *Peary* would be theirs to rip and mangle....

A dragging sound brought Ken's eyes wearily up and to the side. One of the crew who had been lying on the deck was dragging his body painfully toward a row of lockers at one side of the compartment. The man's eyes were feverishly intent on the lockers.

Ken watched his progress dully, without thinking, as inch by inch he forced himself through the other bodies sprawled in his way. He saw him reach the lockers, and for a minute, gasping, lie there. He saw a clawing arm stretch almost up to the catch on one locker, while the man whimpered like a child at his lack of quick success.

Crash! The grinding blow of the torpoon hitting the quarsteel clanged out from behind. But Ken's mind was all on the reaching man's strange actions. He saw the fingers at last succeed in touching the catch. The door of the locker opened outward, and eagerly the man reached inside and pulled. With a thump, a row of heavy objects strung together rolled out onto the deck—and Ken Torrance sprang suddenly to the man's side:

"What are you doing?" he cried.

The man looked up sullenly. He mumbled:

"Damn fish—won't get me. I'll blow us all to hell, first!"

At that the connection struck Ken.

"Then that's nitromite!" he shouted. "That's the idea—the nitromite!"

And stooping down, he wrenched the rope of small black boxes which contained the explosive from the man who had worked so painfully to get them.

"I'll do the blowing, boy!" he said. "Don't worry; I'll do it complete!"

Ken, holding the rope of explosives, crossed the deck and pulled Sallorsen and Lawson around. Their worn faces, with lifeless, bloodshot eyes, met his own strong features, and he said forcefully:

"Now listen! I need your help. I've found our one last chance for life. We three are the strongest, and we've got to work like hell. Understand?"

His enthusiasm and the vigor of his words roused them.

"Yes," said Lawson. "What—we do?"

"You say there's an hour's air left in the sea-suits?" Torrance asked the captain.

"Yes. An hour."

"Then get the men into the suits," the torpooner ordered. "Help the weaker ones; slap them till they obey you!" There came the ugly, deafening crash of the hurled torpoon into the compartment door. Ken finished grimly: "And for God's sake, hurry! I'll explain later."

Sallorsen and Lawson unquestioningly obeyed. Ken had reached the spirit in them, the strength not physical, that had all but been driven out by the long, hopeless weeks and the poisonous stuff that passed for air, and it had risen and was responding. Sallorsen's voice, for the first time in days, had his old stern tone of command in it as, calling on everything within him, he shouted:

"Men, there's still a chance! Everyone into sea-suits! Quick!"

A few of the blue-skinned figures lying panting on the deck looked up. Fewer moved. They did not at once understand. Only four or five dragged themselves with pathetic eagerness towards the pile of sea-suits and the little store of fresh air that remained in them. Sallorsen repeated his command.

"Hurry! Men—you, Hartley and Robson and Carroll—your suits on! There's air in them! *Put 'em on!*"

And then Lawson was among them, shaking the hopeless, dying forms, rousing them to the chance for life. Several more crawled to obey. By the time the next crash of the torpoon came, eleven out of the twenty-one survivors were working with clumsy, eager fingers at their sea-suits, pushing feet and legs in, drawing the tough fabric up over their bodies, sliding their arms in, and struggling with quick panting breaths to raise the heavy helmets and fasten them into place. Then—air!

Again the ear-shattering crash. The scientist and the captain drove at the rest of the crew. They stumbled, those two fighting men, and twice Lawson went down in a heap as his legs gave under him; but he got up again, and they began dragging the suits to the men who had not even the strength to rise, shoving inert limbs into place, switching on the air-units inside the helmets and, gasping themselves, fastening the helmets down. Theirs was a conflict as cruel, as hard and brutal as men smashing at each other with fists, and they then proved their right to the shining roll of honor, wherever and whatever that roll may be. They fought on past pain, past sickness, past poisoning, that man of action and men of the laboratory.

And outside that foul transparent pit the tempo quickened also. The sledging blows at the last door came quicker. All around the captive *Peary* the sleek brown bodies stirred uneasily. For weeks there had been but little activity inside the submarine; now, all at once, three of the figures that were men whipped the others into action, rousing those lying dying on the deck—working, working. Observing this, the lithe seal bodies moved with new nervous, restless strokes, to and fro, never pausing—passing up and down in a milling stream the length of the craft, clustering closest outside the walls of the fourth compartment, where they pressed as close as they could, their wide brown eyes already on the haggard forms that worked inside, their smooth bodies patterned by the constantly shifting shadows of their fellows above and behind.

So they watched and waited, while in the third compartment the battered torpoon was slung at the last door, and drawn back, and slung again—waited for the final moment, the crisis of their month-long siege beneath the floes of the silent Arctic sea!

Kenneth Torrance worked by himself.

He saw that Sallorsen and Lawson had answered his call; man after man was clad in his suit and sucking in the incomparably fresher, though artificial, air of the units. As he had hoped, that air was revitalizing the worn-out bodies rapidly, giving them new strength and clearing their brains. His plan required that—strength for the men to move and act for themselves—sane heads!

The plan was basically simple. Bringing his best concentration to the all-important details, Ken started to build the road to the world above.

First he opened the inner door of the starboard port-lock, wherein lay his torpoon. Opening the entrance panel of the steel shell, he quickly transferred within the cans of compressed food retrieved from the second compartment. When he had finished, there was left barely room for the pilot's body.

And then the nitromite.

The explosive was carried by the *Peary* for the blasting of such ice floes as might trap her. It was contained for chemical stability in a half dozen six-inch-square, water-proof boxes, strung one after another on an interconnecting wired rope. Ken would need them all; he wished he had five times as many. It would not matter if the whole of the *Peary* were shattered to slivers.

Ken tied the rope of boxes into a strong unit, as small as it could be made. Firing and timing mechanisms were contained in each unit: he would only have to set one of them. He wrapped the whole charge, except for one small corner, in several pieces of the men's discarded clothing—monkey jackets, thick sweaters, a dirty towel—and stuffed it in an empty tin container for sea-biscuits.

All this had taken only minutes. But in those minutes the quarsteel of the watertight door had been subjected to half a dozen smashing blows, and already a flaw had appeared in the pane. Another grinding crunch, and there would be the visible beginning of a crack. Three more, perhaps, and the door would be down.

But the plan was laid, the counter move ready; and, as Sallorsen and Lawson, last of them all, got into suits, Ken Torrance, in short, gasping sentences, explained it.

"All the nitromite's in this," Ken said. "I hope it's enough. In a moment I'll set the timing to explode it in one minute—then eject it from the empty torpoon port-lock. It's a gamble, but I think the explosion should kill every damned seal around the sub. Water carries such shocks for miles, so it should stun, if not kill, all the others within a long radius. See? We're inside sub, largely protected. When the stuff explodes, you and men make for the hole you blew in the ice above."

Another crash sent echoes resounding through the remaining compartment. All around the three were suit-clad figures, grotesque clumsy giants, all feeling new strength as they gulped with leathern throats and lungs at the artificial air which was giving them a respite, however brief, from the death they had been sinking into. In the third compartment of the *Peary*, five seal-like creatures with swift and beautiful movements picked up their torpoon battering ram again; while all around the outside of the *Peary* their hundreds of watching fellows pressed in closely.

"Yes!" cried Lawson, the scientist. "But the explosion—it might shatter the ship!"

"No matter; I expect it to!" answered Ken. "Then you can leave through a crack instead of a port-lock."

"Yes—but you!" objected the captain. "Get on a suit!"

"No; I'm jumping into my torpoon in the other port-lock. I've got the food in it. Now, Sallorsen, this is your job. I'll be in my torpoon, but I won't be able to let myself out the port. You open it, right after the explosion. Understand?"

"Yes," replied Sallorsen, and Lawson nodded.

"All right," gasped Ken Torrance. "Empty the chamber." As the captain did so, Ken opened the lid of the biscuit can and adjusted the timing device on the exposed unit in the clothing-wrapped bundle. Then he replaced it, ticking, in the can and thrust the can bodily into the emptied chamber of the port-lock. He closed the inner door of the chamber, and said to the men by him:

"Close your face-plates!"

And Ken pushed the release button: and then he was running to the other port-lock and to his torpoon, and harnessing himself in.

His brain teemed with the possibilities of the situation as he lay stretched out in the torpoon, waiting. How much would the submarine be smashed? Would the charge of nitromite, besides killing the sealmen, kill everyone inside the *Peary*? For that matter, would it affect the sealmen at all? How much could the creatures stand? And would the firing mechanism work? And then would he himself be able to get out; or would the lock in which the torpoon lay be damaged by the explosion and trap him there?

Seconds, only seconds, to wait, small fractions of time—but they were more important than the days and the weeks that the *Peary* had lain, a lashed-down captive, under the Arctic ice; for in these seconds was to be given fate's final answer to the prayer and courage of them all.

Time for Ken expanded. Surely the charge should have gone off long before this! The pulse beat so loudly in his brain that he could hear nothing else. He counted: "... nine, ten, eleven—" Had the fuse failed? Surely by now—"... twelve, thirteen, fourteen—"

On that the submarine *Peary* leaped. Ken Torrance, himself inside the torpoon, felt a sharp roll of thunder made tangible, and then complete darkness took him....

CHAPTER VII
THE AWAKENING

He had no idea of how long he had been unconscious when, his full senses returning, he eagerly peered ahead through the torpoon's vision-plate. For some seconds he could see nothing; but he knew, at least, that the torpoon had survived the shock, for he was dry and snug in his harness. And then his eyes became accustomed to the darkness, and he saw that he was outside the submarine. Sallorsen had followed his orders; had opened the port-lock! The undersea reaches lay ahead of him, and the way was clear.

Ken stared into a gray, silent sea, no longer shadowed with moving brown-skinned bodies. He tried his motors. Their friendly, rhythmic hum answered him, and carefully he slipped into gear and crept up off the sea-floor. He did not dare use his lights.

The *Peary* was a great, blurred shadow, a dead thing without glow or movement, with no figures of sealmen around her. As Ken's eyes gained greater vision, he was able to make out a wide, long rent running clear across the top of the fourth compartment of the submarine. The explosion had done that to her, but what had it done to her crew? What had it done to the sealmen?

He saw the sealmen first. Some were quite close, but in the murk he had missed them. Silent specters, they were apparently lifeless, strewn all around at different levels, and most of them floating slowly up toward the dim ice ceiling.

But up under the ice was movement! Living figures were there! And at the sight Kenneth Torrance's lips spread in their first real grin for days. The plan had worked! The sealmen had been destroyed, and already some of the *Peary's* men were up there and fumbling clumsily across the hundred feet which separated them from the hole in the ice that was the last step to the world above.

A ghostly gray haze of light filtered downward through the water from the hole. Ken counted twelve figures making their way to it. As he wondered about the rest of the crew, he saw three bulging, swaying shapes suddenly emerge from the split in the top of the *Peary*, and begin an easy rise toward the ice ceiling ninety feet above. There was no apparent danger, and they went up quite slowly, with occasional brief pauses to avoid the risk of the bends. Clasped together, the group of three were, and when they were halfway to the glassy ceiling of the ice, three more left the rent in the submarine and followed likewise. Twelve men were at the top; six others were swimming up; three more were yet to leave the submarine—and after they had abandoned her, he, Ken, would follow with the torpoon and the food it contained.

So he thought, watching from where he lay, down below, and there was in him a great weariness after the triumph so bitterly fought for had been achieved. He rested through minutes of quiet and relaxation, watching what he had brought about; but only minutes—for suddenly without warning all security was gone.

From out the murky shadows to the left a sleek shape came flashing with great speed, to jerk Ken Torrance's eyes around and to widen them with quick alarm.

A sealman! A sealman alive, and moving—and vengeful! A sealman which the explosion of nitromite had not reached!

Doubtless the lone creature was surprised upon seeing all its fellows motionless, drifting like corpses upward, and the men of the *Peary* escaping. With graceful, beautiful speed, a liquid streak, it flashed into the scene, eeling up and around and down, trying to understand what extraordinary thing had happened. But finally it slowed down and hovered some thirty feet directly above the dark hull of the *Peary*.

The men rising toward the ice had seen the sealman at the same time Ken Torrance had, and at once increased their efforts, fearing immediate attack. Quickly the two groups shot to the top where the other twelve were, and began a desperate fumbling progress over toward the hole that alone gave exit. But the sealman paid no attention to them. It was looking at something below.

Ken saw what it was.

The last three men were leaving the *Peary*. Awkward, swaying objects, they rose up directly in front of the hovering creature.

With an enraged thrust of flippers, it drove at them. The three humans—Sallorsen, Lawson and one other, Ken knew they must be—were clasped together, and the long, lithe, muscular body smote them squarely, sent them whirling and helpless in different directions in the sea-gloom. One of them was driven down by the force of the blow, and that one the sealman chose to finish first. It lashed at him, its strong teeth bared to rip the sea-suit, concentrating on him all the rage and all the thirst for vengeance it had.

But by then, down below, the torpoon's motors were throbbing at full power; the thin directional rudders were slanting; the torpoon was turning and pointing its nose upward; and Ken Torrance, his face bleak as the Arctic ice, was grasping the trigger of the nitro-shell gun.

He might perhaps have saved the doomed man had he swept straight up then and fired, but a quick mounting of the odds distracted him for a fatal second. Out of the deeper gloom at the left came a swiftly growing shadow, and Ken, with a sinking in his stomach, knew it for a second sealman.

Then another similar shadow brought his eyes to the right.

Two more sealmen! Three now—and how many more might come?

At once Ken knew what he must do before ever he fired a shell at one of the brown-skinned shapes. The man just attacked had to be sacrificed in the interests of the rest. The torpoon swerved, thrust up toward the ice ceiling under the full force of her motors; and when halfway to it, and her gun-containing bow was pointed at a spot in the ice only twenty feet in front of the foremost of the men stroking desperately towards the distant exit-hole, Ken pressed the trigger; and again, and again and again....

Twelve shells, quick, on the same path, bit into the ice. Almost immediately came the first explosion. It was swelled by the others. The ice shivered and crumbled in jagged splinters—and then there was a new column of light reaching down from the world of air and life into the darkness of the undersea. A roughly circular hole gaped in the ice sixty or seventy feet nearer the swimming men than the old one.

"That'll give 'em a chance," muttered Kenneth Torrance. He plunged the torpoon around and down. "And now for a fight!"

Without pause, now, there was, straight ahead, a hard, desperate duel, a fitting last fight for any torpoon or any man riding one. Each of the seven shells left in the nitro-gun's magazine had to count; and the first of them gave a good example.

Ken turned down in time to see the death of the man first attacked. His suit was ripped clean across, his air of life went up in bubbles, and the water came in. The seal-creature lunged at its falling victim a last time, and as it did so its smooth brown body crossed Ken's sights. The torpooner fired, and saw his shell strike home, for the body shuddered, convulsed, and the sealman, internally torn, went sinking in a dark cloud after the human it had slain.

That sight gave pause to the other two creatures that had arrived, and gave Ken Torrance a good second chance. Motor throbbing, the torpoon turned like a thing alive. Its snout and gun-sights swerving straight toward the next target. But, when just on the point of pressing the trigger, Ken's torpoon was struck a terrific blow and tumbled over and over. The whole external scene blurred to him, and only after a moment was he able to bring the torpoon back to an even keel.

He saw what had happened. While he had been sighting on the second seal-creature, the third had attacked the torpoon from the rear by striking it with all the strength of its heavy, muscular body. But it did not follow up its attack. For it had crashed in to the whirling propeller, and now it was hanging well back, its head horribly gashed by the steel blades.

For a moment the three combatants hung still, both sealmen staring at the torpoon as if in wonder that it could strike both with its bow and stern, and Ken Torrance rapidly glancing over the situation. The remaining two of the last group of three men, he saw, had reached the top, and the foremost of the *Peary's* crew were within several feet of the new hole in the ice. In a very short time all would be out and safe. Until then he had to hold off the two sealmen.

Two? There were no longer only two, but five—ten—a dozen—and more. The dead were coming to life!

Here and there in the various levels of drifting, motionless brown bodies that he thought the explosion had killed, one was stirring, awakening! The explosion had but stunned many or most of them, *and now they were returning to consciousness*!

CHAPTER VIII
THE DUEL

Upon seeing this, all hope for life left Ken. He had only six shells left, and at best he could kill only six sealmen. Already, there were more than twenty about him, completely encircling the torpoon. They seemed afraid of it, and yet desirous of finishing it—they hung back, watching warily the thing that could strike and hurt from either end; but Ken knew, of course, that he could not count on their inaction long. One concerted charge would mean his quick end, and the death of most of the men above.

Well, there was only one thing to do—try to hold them off until those men above had climbed out, every one.

With this plan in mind, he maneuvered for a commanding position. Quietly he slid his motor into gear, and slowly the torpoon rose. At this first movement, the wall of hesitating brown bodies broke back a little. It quickly pressed in again, however, as the torpoon came to a halt where Ken wanted it—a position thirty feet beneath, and slightly to one side, of the escaping men above, with an angle of fire commanding the area the sealmen would have to cross to attack them.

Almost at once came action. One of the surrounding creatures swerved suddenly up toward the men. Instinctively angling the torp, Ken sent a nitro-shell at it; and the chance aim was good. The projectile caught the sealman squarely, and, after the convulsion, it began to drift downward, its body torn apart.

"That'll teach you, damn you!" Ken muttered savagely, and, to heighten the effect he had created, he brought his sights to bear on another sealman in the circle around him—and fired and killed.

This sight of sudden death told on the others. They grew obviously more fearful and gave back, though still forming a solid circle around the torpoon. The circle was ever thickening and deepening downward as more of those that the explosion had rendered unconscious returned to life.

And then, above, the first man reached the hole, clawed at its rough edges and levered himself through.

That was a signal. From somewhere beneath, two brown bodies flashed upward in attack. Fearing a general rush at any second, Ken fired twice swiftly. One shell missed, but the other slid to its mark. Almost alongside its fellow, one of the creatures was shattered and torn, and that evidently altered the other's intentions, for it abandoned the attack and sought safety in the mass of its fellows on the farther side.

Another respite. Another man through the hole. And but two nitro-shells left!

The deadly circle, like wolves around a lone trapper who crouches close to his dying fire, pressed in a little; and by their ominous quietness, by the sight of their eyes all turned in on him, their concerted inching closer, Ken sensed the nearness of the charge that would finish him. All this in deep silence, there in the gloomy quarter-light. He could not yell and brandish his fists at them as the trapper by the fire might have done to win a few extra minutes. The only cards he had to play were two shells—and one was needed now!

He fired it with deliberate, sure aim, and grunted as he saw its victim convulse and die, with dark blood streaming. Again the swarm hesitated.

Ken risked a glance above. Only three men left, he saw; and one was pulled through the hole as he watched. Below, in one place, several seal-creatures surged upward.

"Get back, damn you!" he cursed harshly. "All right—take it! That's the last!"

And the last shell hissed out from the gun even as the last man, above, was pulled through up into the air and safety.

Ken felt that he had given half his life with that final shell. Completely surrounded by a hundred or more of the sealmen, he could not possibly hope to maneuver the torpoon up to the hole in the ice and leave it, without being overwhelmed. He had held off the swarm long enough for the others to escape, but for himself it was the end.

So he thought, and wondered just when that end would come. Soon, he knew. It would not take them long to overcome their fear when they saw that he no longer reached out and struck them down in sudden bloody death. Now it was their turn.

"Anyway," the torpooner murmured, "I got 'em out. I saved them."

But had he? Suddenly his mind turned up a dreadful thought. He had saved them from the sealmen, but they were up on the ice without food. There had been no time to apportion rations in the submarine; all the supplies were stacked around him in the torpoon!

Searching planes would eventually appear overhead, but if he could not get the food up to the men it meant their death as surely as if they had stayed locked in the *Peary*!

But how could he do it without shells, and with that living wall edging inch by inch upon him, visibly on the brink of rushing him. Some carried ropes with which they would lash the torpoon down as they had the others. Must all he and those men had gone through, be in vain? Must he die—and the others? For certainly without food, those men above on the lonely ice fields, all of them weakened by the long siege in the submarine, would perish quickly....

And then a faintly possible plan came to him. It involved an attempt to bluff the seal-creatures.

Thirty feet above the lone man in the torpoon was the hole he had blasted in the ice. He knew that from the cone of light which filtered down; he did not dare to take his eyes for a second from the creatures around him, for all now depended on his judging to a fraction just when the lithe, living wall would leap to overwhelm him.

Now the torpoon was enclosed by what was more a sphere of brown bodies than a circle. But it was not a solid sphere. It stretched thinly to within a few feet of the ice ceiling where, in one place, was the hole Ken had blown in the ice.

He began to play the game. He edged the gears into reverse, gently angled the diving-planes, and slowly the torpoon tilted in response and began to sink back to the dark sea-floor.

Motion appeared in the curved facade of sleek brown heads and bodies in front and to the sides. The creatures behind and below, Ken could not see; he could only trust to the fear inspired by the damage his propeller had wreaked on one of them, to hold them back. However, he could judge the movements of those behind and below by the synchronized movements of those in front; for the sealmen, in this tense siege, seemed to move as one—just as they would move as one when a leader got the courage to charge across the gap to the torpoon.

In reverse, slowly, the torpoon backed downward. Every minute seemed a separate eternity of time, for Ken dared not move fast at this juncture, and he needed to retreat not less than fifty feet.

Fifty feet! Would they hold off long enough for him to make it?

Foot by foot the torpoon edged down at her forty-five-degree angle, and with every foot the watching bodies became visibly bolder. There was no light inside the torpoon—inner light would decrease the visibility outside—but Ken knew her controls as does the musician his instrument. Slowly the propeller whirled over, the torpoon dropped, slowly the diffused light from the hole above diminished—and slowly the eager wall of sealmen followed and crept in.

Twenty-five feet down; and then, after a long time, thirty-five feet, and forty. Seventy feet up, in all, to the hole in the ice....

Ken wanted seventy-five feet, but he could not have it. For the wall of sleek bodies broke. One or two of the creatures surged forward; other followed; they were coming!

The slim torpoon leaped under the unleashed power of her motors—forward.

For one awful moment Ken thought he was finished. The vision of the hole was obscured by a twisting, whirling maelstrom of bodies, and the torpoon quivered and shook like a living thing in agony under glancing blows.

But then came a patch of light, a pathway of light, leading straight up at a forty-five-degree angle to the hole in the ice above.

Sealmen and torpoon had leaped forward at the same moment. Doubtless the creatures had not expected the shell to move so suddenly and decisively ahead, so that when it did, those in the van swerved to escape head-on contact.

The torpoon gained speed all too slowly for her pilot. It naturally took time to gain full forward speed from a standing start. But she moved, and she moved fast, and after her poured the full tide of sealmen, now that they saw their prey running in retreat.

From somewhere ahead appeared a rope, noosed to catch the fleeing prey. It slipped off the side. Another touched the bow, but it too was thrown off. The torpoon's forward momentum was now great; she was sweeping up at the full speed Ken had gone back to be able to attain. He needed full speed! The plan would fail at the last moment without it!

Another rope; but it was the seal-creature's last gesture. Through the side plates of quarsteel the light grew fast; the ice was only ten feet away; a slight directional correction brought the hole dead ahead—and at full speed, twenty-four miles an hour, the torpoon passed through and into the thin air of the world of light and life.

Right out of the hole, a desperate fugitive from below, she leaped, her propeller suddenly screaming, and arched high through the air before she dove with a rending, splintering crash onto the upper side of the sheet ice.

And the sun of a cloudless, perfect Arctic day beat down on her; and men were all around, eagerly reaching to open her entrance port. It was done.

Kenneth Torrance, dazed, battered, hurting in every joint but conscious, found the torpoon's port open, and felt hands reach in and clasp him. Wearily he helped them lift him out into the thin sunlight. Sitting down, slitting his eyes against the sudden glare, he peered around.

Captain Sallorsen was beside him, supporting him with one hand and pounding him on the back with the other; and there in front was the bearded scientist, Lawson, and the rest of the men.

Ken took a great gulp of the clean, cold air.

"Gosh!" was all he could say. "Gosh, that tastes good!"

"Man, you did it!" shouted Sallorsen. "How, in God's name, I don't know—but you did it!"

"He did!" said Lawson. "And he did it all himself. Even to the food, which should keep us till a plane comes by. If they haven't stopped searching for us."

His words reminded Ken of something.

"Oh, there'll be a plane over," he said. "Forgot to tell you, but I stole this torpoon—see?—and told the fellows they could come and get it somewhere right around here."

Kenneth Torrance grinned, and glanced down at the battered steel shell which had borne him out of the water below.

"And here it is," he finished. "A little damaged—but then I didn't promise it would be as good as new!"